A Walk through *A World of Plants:*
The Enid A. Haupt Conservatory

A WALK THROUGH

A World of Plants

By Allan Appel

Christine M. Douglas, Art Editor

THE ENID A. HAUPT CONSERVATORY
AT THE NEW YORK BOTANICAL GARDEN

Published by The New York Botanical Garden
Bronx, New York U.S.A.

The paper used in this publication meets the requirements
of the American National Standard for Information Sciences—
Permanence of Paper for Publications and Documents in
Libraries and Archives.
ANSI/NISO (Z39.48-1992)

Printed in the United States of America using
soy-based ink on acid-free recycled paper.

Library of Congress Cataloging-in-Publication Data

Appel, Allan.
    A walk through a world of plants: the Enid A. Haupt
    Conservatory; by Allan Appel.
    p. cm.
    ISBN 0-89327-425-9 (alk. paper)
    1. Enid A. Haupt Conservatory—Guidebooks.
    I. New York Botanical Garden.
QK73.U62N492   1996
580'.7'3747275—DC21
                                          96-54479
                                          CIP

ISBN 0-89327-425-9

This book is dedicated to

Enid Annenberg Haupt

in recognition of her love
for The New York Botanical Garden,
the plants that live here,
and the people who care for them.

CONTENTS

THE GREAT CONSERVATORY: ALIVE AGAIN

After four years of restoration, the Enid A. Haupt Conservatory, home to *A World of Plants,* opened May 1, 1997. The Board of The New York Botanical Garden wishes to recognize the astounding philanthropic and creative contributions of the following human beings without whom this project would not have been possible:

Enid A. Haupt, *Garden Board Member*
John Belle and Richard Southwick, *Architects*
Jon Coe, *Exhibition Designer*
Mrs. Donald B. Straus, *Senior Vice Chairman*
John Rorer, *Executive Vice President*
Richard Schnall, *Vice President for Horticulture*
Leslie Loring, *Director of Government Relations*
Amy Cohn, *Director of Capital Projects*
Joseph Kerwin, *Conservatory Manager*
Francisca Coelho, *Curator of Tropical Plants*
Ernest DeMarie, Ph.D., *Curator of Desert Plants*
Bill Fitzgerald, *Construction Manager*
Charles Frattini, *Construction Superintendent*

And we wish to thank the officials of New York City responsible for major City funding for the project: Mayor Rudolph W. Giuliani; Speaker of the City Council Peter F. Vallone; Commissioner of Cultural Affairs Schuyler G. Chapin; and Bronx Borough President Fernando Ferrer.

In an inspired collaborative effort, hundreds of other persons—Garden Board members, horticulturists, scientists, architects, designers, engineers, financial planners, government relations specialists, fundraisers, writers, photographers, administrators, and members of the construction trades have worked together on this massive $25,000,000 project every day since late 1988. They cannot all be named here, unfortunately, but to all of them, congratulations on a task magnificently accomplished.

Thomas J. Hubbard   Gregory Long
Chairman   *President*

A JOURNEY BEGINS

...a body corporate by the name of The New York
Botanical Garden...for the advancement of botanical
science...exhibition of horticulture...and for the enter-
tainment, recreation, and instruction of the people.
from the Charter of The New York Botanical Garden, 1891

Welcome

In stepping into the Enid A. Haupt Conservatory of The
New York Botanical Garden, you are about to begin a
grand botanical adventure through the beautiful, exotic,
and—most significantly—indispensable realm of living
plants. You will look up at species of *Roystonea, Euterpe,*
and *Acrocomia* palms in the Palms of the Americas gallery,

Zamia inermis.

where your journey begins. Using your imagination as you walk through the Conservatory—as we hope you will—you will be able to transport yourself not only to faraway rain forests, deserts, and mountain slopes, but you might also move through time as well. This is because many plants such as the cycads are survivors of great evolutionary changes. Their leaves, which may graze your arm as you walk by, are not that different today, so the fossil record reveals to us, from the leaves of their ancestors geological ages ago.

Rain Forests

Many sites are scheduled on your trip: the lush Tropical Lowland Rain Forest, where you will be able to climb a skywalk and enter a vast green canopy of trees that, in the real rain forest, are so tightly interwoven that only a tiny percentage of sunlight filters down to the forest below. Then you'll be off to the misty mountains of the Tropical Upland Rain Forest, where orchids and other epiphytes— plants that live on other plants—grow in multi-colored profusion. The world's rain forests, which these displays will evoke for you, are the most species-rich environments in the world. In them we see most clearly and intensely nature's greatest living experimental laboratory in plant and animal adaptation, interdependence, and survival.

Scientific Research

An important part of your trip will be a visit to a Research Station of the sort that New York Botanical Garden scientists maintain in the field; also, you'll see an indigenous peoples' Healer's House and Garden. At these locations you may be able to imagine the site where drugs to treat cancer or AIDS may be found—because only a very small fraction of rain forest plants have been fully studied for their medicinal potential. Of course, for such discoveries to occur, it is fundamental that the rate of rain forest destruction be slowed and then halted. At the Conservatory's Terrace Gardening site, you'll see how farmers work step-like plots on steep rain forest slopes to keep the soil from washing away. Garden scientists work with local people, whose management of land and crops is varied and innovative. Local farmers are often able to take advantage of new market opportunities without destroying the rain forests and their irreplaceable richness of species—their biodiversity. Studying ways to slow the destruction of the rain forest is an important research topic for Garden scientists.

Deserts

After the rain forests, you journey to the Deserts of the World. Far from "deserted," these environments are rich in unusual plant species. In the saguaro cactus, the agaves, and the aloes, you will see the seemingly infinite variety of plant adaptations to a profoundly dry world. Beyond the deserts are stops among the Garden's collections of carnivorous plants (when last checked they were still digesting

A cruciform house and Magnolia grandiflora.

insects only, not people), plants from the Subtropics, and the Garden's world-famous horticultural displays.

The "botanical vessel" making your journey possible is the Enid A. Haupt Conservatory, one of the most significant historic glasshouses left in America. First opened in 1900, it is designated both as an official New York City Landmark and, along with the entire Garden, as a National Historic Landmark. At last restored to its full architectural integrity and glory, the Conservatory has been fitted with modern computer-controlled misting and temperature control systems. Symbolic of the Garden and its mission, the Conservatory and its exhibition, *A World of Plants,* are powered by a twenty-first-century sense of urgency. Because extinction is final, the loss of our botanical and zoological heritage is the greatest of all the serious environmental crises facing us. That is why, to an extent unparalleled in the history of the Conservatory, visitors can now venture to the plants themselves in re-creations of their natural rain forest or desert context—what scientists call biomes.

On this journey you will learn not only about plants' remarkable processes of pollination, reproduction, and adaptation, but also about their relationship to soil, water, light, and climate and their vital and complex connection to humankind. Your guide, in part, will be the accumulated insights of the Garden's scientists, whose pioneering, extensive, and preeminent studies in the field and in the laboratory over the last century have often focused on rain forests and deserts, where new scientific data and new environmental policies based upon them are most urgently needed.

Life on Earth

Without green plants, it is unlikely that there would be any other life on earth; without them and the remarkable process of photosynthesis, which no scientist has been able to replicate in the laboratory, where would we get the oxygen in the air? Where would we obtain the foods that sustain us; the fuels that warm us; the roofs that shelter us; the wines for celebrating; or the medicines for treating pain and diseases today and in the future?

This is why it is fundamental to explore plants and the relationship between plants and people, and you will discover both on your travels through the Conservatory. Without plants we could not survive. So breathe deeply, thank the plants, and enjoy your journey.

PALMS OF
THE AMERICAS

Man *dwells* naturally within the tropics and
lives on the fruit of the palm tree. He *exists* in
other parts of the world and there makes shift
to feed on corn and flesh.

— *Carolus Linnaeus (1707-1778),*
founder of modern scientific
classification of plants and animals

In beginning your visit to the Conservatory, you have
stepped into the world's most extensive greenhouse collection of palms from the Americas—both North and South.
With 2,500 species of palm existing worldwide, and 550
species identified from the American tropics thus far, palms
are as diverse as they are useful to millions of people
around the world.

Indeed, palms provide an extraordinary array of uses
wherever they are found: trunks and stems for construction materials, bows, and arrowheads; leaves for roofs,
brooms, and food; fruits for beverages, food, and fuel; and
oils for cooking and in margarine, candies, soaps, cosmet-
ics, and engine lubricants. Even the hard seeds of some
species are fashioned into shiny buttons and ornaments. In
their great diversity, utility, and ability to grow even in
poor soils, palms have been associated intimately with the
development of civilization.

In the Americas, durable and well-traveled species, such
as the coconut palm, have colonized whole tropical coast-
lines and islands. Over generations, palms have become so
central to life and the economy in some communities that
they have taken on a religious significance. Yet the contributions of palms, many of whose habitats are under siege,

The trunk of a
Roystonea regia,
royal palm, with other
American palms in
the background
surrounding the pool.

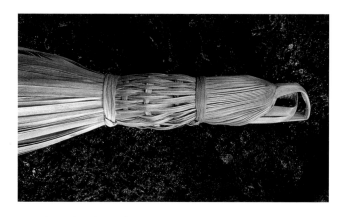

A broom handle made from a Coccothrinax *palm leaf in the Caribbean.*

remain seriously under-appreciated by people in the developed world.

Palms: Princes, Trees of Life

Many species of palm grow in lush tropical rain forests. In fact 75 percent of known species favor the warm, moist conditions found there. The root systems of palms, however, can adapt to different environmental conditions. The buriti palm, *Mauritia flexuosa,* grows in wet grassy areas. The sturdy thatch palm, *Sabal yapa,* is tough and versatile, flourishing in dry limestone soil.

Products from the Sabal yapa *palm.*

Palms are so widespread, diverse, and—as you will see if you study the coconut palm more closely—so dependable a source of goods and services to human society that none other than Carolus Linnaeus, the eighteenth-century Swedish founder of modern botanical classification, anointed them "princes" of the plant world.

Palmae: Systematic Botany

Wise use of plants, conservation of biological diversity, and a full and meaningful appreciation of the world's resources all depend ultimately on a body of systematic fact.

—Rupert C. Barneby, Ph. D.
Curator, Institute of Systematic Botany,
The New York Botanical Garden

All palms—indeed, all known plants—have been classified by Linnaeus and the generations of "systematic botanists" who have followed. This body of scholars includes scientists at the Garden such as Andrew Henderson, Ph.D., one of the Garden's palm experts. When Dr. Henderson collects or identifies a new species—and he has discovered seven species of palm new to science—he enters it into the Herbarium, that vast databank of systematic botany. This science studies the evolutionary links among species by gathering information about their traits, including seed form and size. Then by synthesizing these data into encyclopedic monographs and floristic treatises, systematic botanists ultimately formulate hypotheses that are the best statements of relationships among plant groups that modern science can provide. Systematic botany, the Garden's central research area, is a science that provides a rational basis for the conservation and wise utilization of plant resources worldwide.

The classification system works this way: Species are the smallest groups of organisms that are consistently and persistently distinct from other such groups and distinguishable by ordinary means. Related species constitute a genus; related genera (plural of genus) form a family, and so on from family to order to class. The palms, known by their Latin family name, Palmae, are a big beautiful family indeed.

Among key characteristics scientists look at in studying

Herbarium specimens,
Raphia taedigera *(left)*
and Socratea exorrhiza
(right), were collected by
Andrew Henderson,
Ph.D., on expeditions to
the tropics.

Trunk of a Roystonea
oleracea *royal palm.*

The pinnate leaf of
Acrocomia aculeata.

22

and grouping palms are trunk shape—notice the slender cylindrical trunk of the *Euterpe* palm on the west side of the gallery—and, of course, the leaves, flowers, and fruits. Palms have either palmate (hand or fan-like) leaves such as those of the wax palm (*Copernicia prunifera*) or pinnate (feather-like) leaves such as those of the royal palm (*Roystonea oleracea*). Notice the set of five royal palms, native to Cuba, surrounding and reflected in the pool. They may grow to 90 feet, reaching the top of the Conservatory's dome.

The length of palm leaves varies dramatically, and the longest comes from an African species of *Raphia* that may grow to a length of 75 feet. The only Western Hemisphere representative of this genus, *Raphia taedigera,* occurs in swampy areas of the Amazon, and its leaves may grow to 30 feet. This species can be seen in our Tropical Lowland Rain Forest biome. Palms also vary greatly in height from the six-inch lilliput palm, a species of the genus *Butia* of Paraguay, to the wax palms of the Andes, of the genus *Ceroxylon,* whose trunks, were they allowed to rise in the Conservatory to full growth, would lift the 90-foot dome well into the sky.

Ceroxylon quindiuense.

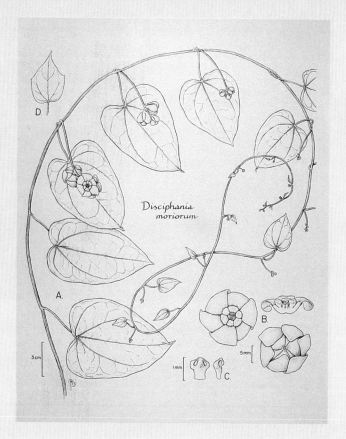

This species was collected for the first time in 1994 on a Garden expedition to French Guiana by Scott Mori, Ph.D., and Carol Gracie. It was described as new to science by Rupert Barneby, Ph.D., drawn by Bobbi Angell, and published in Brittonia *48(1), page 21.*

PLANT EXPLORATION

by Brian M. Boom, Ph.D.
Vice President for Botanical Science
and Pfizer Curator of Botany

Field exploration is the first step in the process of understanding the traits of plants, learning how they are related to each other, determining their names, and researching how they might be useful to society. Back at the Garden, plant specimens are studied in the laboratory, making use of the Library's unexcelled resources, and ultimately housed in the Herbarium, now with nearly 6 million accessions. The results are published in scientific journals. Every year botanists describe hundreds of new species for the first time.

The Coconut Palm — Gift to Humankind

Located near the doorway leading out to the Conservatory's courtyard is one of the best known and significant plants on the planet: *Cocos nucifera,* the coconut palm. As you approach, imagine yourself an early explorer nearing a tropical island. Countless coconut palms, having floated over by way of their durable wave-borne fruit, colonized these shores long before humans arrived. So there they are to greet you and satisfy your thirst, hunger, and need for shelter.

In both its profusion and usefulness in providing for a wide range of human material wants, the coconut palm has been, from first historical acquaintance, one of nature's great gifts to humankind. Coconut meat and milk, its most familiar products, feed millions. Within the coconut fruit resides the endosperm, which, when dried, makes copra. The versatile oil that the copra releases when pressed has for millennia been the foundation of cooking for people in the tropics and far beyond.

Look at the coconut plant's leafy fronds or trunk. You will find that every part of this plant can be used. The leaves are fashioned into thatch for roofing; the fiber is used for mats, rope, and twine; the sap for sugar and wine; the trunk for boats and furniture; and the dried nut shells and other parts for musical instruments and ornaments.

And you, too, are among the millions worldwide who use coconut palm products for daily needs. Just read the small print on the wrapper of a bar of candy or soap or on your bottle of shampoo, and you will see that the coconut is a mainstay of the local as well as the export economy of many tropical countries.

Broom made of Cocos nucifera *leaf mid-ribs.*

Economic Botany and the Glorious Palm

Researching plants for their potential as food, fuel, lubricant, and medicine, to name but a few uses, is the discipline scientists call economic botany. It is also an important part of the mission of The New York Botanical Garden, which established the Institute of Economic

Polo Romero harvesting chicle latex from Manilkara zapota *used for the production of chewing gum in Belize.*

Botany in 1980. Based largely on research initiated in the tropics, Garden economic botanists such as Michael Balick, Ph.D., have brought home the urgent message that the economic value of plants, including the prospect of future genetic enhancement of economically important species, often depends on allowing plants to grow as undisturbed as possible; it also depends on enlightened management of plant resources by local experts at the village level.

Most products from tropical palms, with the exception of coconut palms and African oil palms, are harvested from wild rather than cultivated trees from plantations.

A young folded palm leaf.

The Hearts of Palm Story

The considerable value of some palms has led to dangerous commercial over-exploitation of certain species like those that yield palm hearts. Botanically, the palm heart develops at the minute growing point, or meristem, of the trunk, surrounded by successively larger but still unexpanded leaves. At this stage the young leaves are still folded, crisp, tender, and good to eat. Also known as palmito or palm cabbage, palm hearts are a popular salad in tropical countries, and for decades they have been canned and exported to temperate regions. Unfortunately, once a palm stem is cut for palm heart, that stem dies. In species with solitary stems, removing the heart kills the palm. Palm heart extraction has been extremely destructive to certain species. Fortunately, nowadays, species with clustered stems such as *Euterpe oleracea* and *Bactris gasipaes* are more commonly used. Palm heart extraction from these species does not kill the plant and can be sustainable, that is, harvested indefinitely from plantations at a steady rate that does not deplete the resource.

Conservation or "Last Stands"

Most palm species have not been studied thoroughly for their economic potential or brought into cultivation. Therefore, destruction of palm natural habitats by logging or cattle ranching or conversion to agriculture can lead to an irreplaceable loss of species before economic potential has become evident. During massive land clearing, tropical soils are exposed and compacted by machines. Then heavy rains cause soil erosion and, with this, loss of nutrients

from the ecosystem. In the long run, without a variety of other plant species nearby, the pollinators leave or die, and the palms that remain standing often become, like lonely sentries at the corners of fields, literally, "last stands." Already 90 species of palm, including 45 Western Hemisphere species, are in danger of extinction. Nine species previously known to science are now thought to be lost forever.

A threatened stand of Ceroxylon quindiuense *in the Colombian Andes.*

Natural Forest Management

There are, however, many ways to conserve palms and other tropical plants; some of them are age-old, some new. One approach is to set aside particularly diverse or threatened habitats as protected preserves. Many of the Garden's scientists have worked closely with governments and especially forest communities to select areas for protection, to catalog the resources of these preserves, and to help design appropriate plans to ensure that important species are effectively conserved.

A lithograph of a palm grove in Karl Friedrich Philipp von Martius' Flora Brasiliensis, *1840 - 1906, Vol. 1, Pt. 1. The LuEsther T. Mertz Library at The New York Botanical Garden.*

In other tropical zones, Garden scientists have documented the considerable skills of rural people as forest managers. Traditional techniques are frequently intricate and virtually invisible to the untrained foreigner's eye. Nevertheless, forest-dwelling people have increased substantially the economic value of their forests while preserving much of their natural diversity, unlike modern agriculture and commercial forestry.

Agroforestry is a promising middle ground between agriculture and forest preservation. A traditional practice often updated to suit modern needs, agroforestry combines production of agricultural staples with forest management, preserving some of the qualities of natural forests: good protection of soils, complex ecosystem structure, and biological diversity.

In the Amazon rain forest of South America, there are almost as many agroforestry systems as there are rural families. A typical agroforestry system includes some tall trees that may provide firewood or shade, a few intermediate-sized fruit-bearing trees, and plants growing closer to the ground that yield tubers, condiments, or greens. Many of the palms you see in the Conservatory are used widely

in Latin America as components of agroforestry systems, including the peach palm, *Bactris gasipaes;* the graceful Açaí palm, *Euterpe oleracea;* and the babassu palm, *Attalea speciosa,* a species with particularly great potential for reclaiming tropical land seriously damaged by deforestation.

Euterpe oleracea.

In the Beginning: Algae and the Origins of Species

The palm is not only useful and stately, it also has a botanical lineage dating back millions of years. As you travel through this gallery and others, you will also notice plants, such as the cycads, ferns, and mosses, whose evolutionary path can be traced even farther back in time. These species are closely related to plants that filled ancient forests hundreds of millions of years ago.

As you acquaint yourself with these species of "living fossils," it is an appropriate time to ask a fundamental question: Where did all this life on earth come from? The answer can be found in a brief evolutionary history of plants, which is how plants, born in the seas, have adapted to life on land.

The Arrival of Plants on Land

After the earth cooled, life began in the oceans about 3.85 billion years ago. By 500 million years ago, green algae had evolved, and from this group of single-celled organisms all the multitudinous plant forms you see around you are thought to have evolved. How did this happen? The details of the evolutionary process are still being actively studied by scientists, but the broad outline of the mechanism is clear enough. The key is a process called *natural selection*. Here is how it works: Organisms usually produce many more offspring than can survive, and because the resources needed for life, such as nutrients, water, or simply space, are usually limited, competition ensues. Because individual organisms in populations of a species possess varying traits that have arisen through genetic mutation, some individuals will be better adapted than others in any given environmental situation. It is precisely these individuals that will survive to reproduce and thus pass on advantageous traits to the next generation. Therefore, through natural selection only the fittest survive and successful combinations of traits become genetically encoded, and over time these traits become distinguishing features of the organisms possessing them.

About 400 million years ago, several traits independently began to develop that allowed the organisms possessing them to exist and diversify on land, in contrast to the ocean-bound green algae from which they were derived. The first of these was a cuticle (a kind of protective, waterproof coating that prevents plants from drying out in the air), and this allowed for the evolution of such groups as the liverworts and mosses. A later successful innovation, vascular tissues (which allow nutrients and water to be transported internally over relatively long distances from root hairs to branch tips) led to the evolution of such groups as whisk ferns, lycopods, quillworts, spike mosses, horsetails, and the true ferns. About 300 million years ago, development of a structure we now know as the seed (an adaptation that provided, in one tidy unit, protection for the embryo and food storage tissue for nutrition until the developing plant could rely on photosynthesis for food production) led to the evolution of a major group of plants we call gymnosperms, which means, literally, "naked seeds," in reference to the fact that their seeds are not enclosed by any structure.

About 125 million years ago, a novel development occurred among the gymnosperms that gave an important selective advantage to the individual plants possessing it, namely the enclosure of the seeds by a structure we now know as a fruit. This was obviously a major evolutionary development over the gymnosperms, because these enclosed seed plants, known as angiosperms, or flowering plants, successfully diversified and spread to become the most dominant group of plants in the world today. Among the angiosperms two major groups evolved: those with two seed leaves (dicotyledons) and those with one seed leaf (monocotyledons). Most trees and shrubs are "dicots." Plants such as lilies, orchids, palms, and the bamboos, and other grasses, which you can see to the left of the entryway to the Conservatory, are examples of "monocots."

The queen sago:
Cycas circinalis.

Cycads: Living Fossils of the Plant World

Now spend some time, if you haven't already, with the unusual cycads. The queen sago, *Cycas circinalis*, is the tallest of the Garden's cycads, at 20 feet in 1997. Among cycads, this species is the tallest and, in the wild, could reach approximately 45 feet! This specimen lived through the restoration of the Conservatory, in place, as it was too large to be moved.

Unlike palms, many cycads have stiff leaves that sometimes have spines or prickles. Cycads are gymnosperms whose botanical lineage has been traced back 230 million years. They inhabit forests, woodlands, grassy areas, and even rain forests. The many special features of cycads are designed to conserve water; such adaptations enable the plants to flourish in harsh and often dry environments.

While many of the known cycad species are endemic to (found exclusively on) certain continents, one species, the queen sago, is widespread from Australia to East Africa, perhaps because its seed, like the fruit of the coconut palm, is spongy and can float on water.

Helping to figure all this out is Dennis Stevenson, Ph.D., a Garden botanist and the world's leading expert on cycads in the tropics. He studies cycad life cycles and, most recently, has focused on the role of insects in the pollination of these ancient plants. This work is increasingly important because all cycads are relatively rare, with 54 of the 220 known species worldwide already endangered.

Now, do what successful plant species do. Migrate to the first biome on your tour of *A World of Plants* in the Enid A. Haupt Conservatory, the Tropical Lowland Rain Forest.

The atala butterfly (Eumaeus atala), *once considered extinct, owes its survival to the cycad* Zamia *on which it feeds. Examples of* Zamia *are on view in this gallery.*

TROPICAL LOWLAND RAIN FOREST

Evolution's Busiest Laboratory on Earth

The fallen branch of the kapok tree is covered with epiphytes.

In this biome you will see a tangle of lianas—twining woody vines—ever climbing to the sunlight above. Among this profusion is *Theobroma cacao,* a tree that grows to 25 feet and bears the seeds that are made into chocolate. Nearby there will be guava, papaya, and cashew, all culti-vated trees of great economic and medicinal value. Just as in the real rain forest, here you feel the constant, warm, moist temperature and the perpetual presence of rainfall.

From the canopy high above to the forest floor below, imagine yourself moving through a great biological cafeteria, open 24 hours a day and never less than completely filled. The meal-making is continual, with the leftovers and the discarded debris on the forest floor quickly incorporated back into more plants and more animals. This cycle continues everywhere as thousands upon thousands of plant and animal species compete for space, light, nutrients, and water.

Rain forests cover less than seven percent of the planet's land surface. And yet about 90,000 of the approximately 300,000 known species of plants in the world are found in the tropical rain forests of the Americas—an incredible botanical richness. They share intensely close quarters, like a vast evolutionary apartment building rising on every acre. From the floor of the forest, to the understory, to the canopy, to the emergent trees protruding into the rainy sky of the Amazon Basin, each level is inhabited by characteristic species. At every opportunity many species of plants climb for a higher apartment ever closer to the life-giving equatorial sunlight.

Biological Diversity

Ithomid butterfly laying eggs on a Solanum torvum *leaf.*

This concentration of life—this biological diversity, or "biodiversity"—gives experts such as the Garden scientists who study it an unequalled opportunity to learn and then disseminate fundamental ecological lessons. Perhaps the chief lesson is that all species of plant and animal life, including human beings, live by giving to and taking from each other. This interdependence is fragile and depends on preservation of individual species as parts of whole ecosystems—soil, water, climate, the pollinating insects, and the plants themselves, the whole humming interactive environment.

In scores of pioneering field research expeditions to the rain forests during the past 100 years, New York Botanical Garden scientists, in collaboration with scientists from around the world, have provided fundamental data to begin answering these questions: Which plants are out there? Where are they? How are they related to one another? How do they interact with animals? How do they benefit indigenous people? How might they benefit us as well? Without such basic information, local scientists in Ecuador, Brazil, Belize, and elsewhere will be unable to train new generations of botanists and ecologists. These local experts can likewise provide government officials and private land-use managers with the data to help them make informed conservation and development policy decisions.

Morpho butterfly drinking the juice of a fruit of Bellucia.

As you enter, the trees to the left and right of your path are representative of life in the understory and the other fiercely competitive zones of the forest beneath the canopy level. While these specimens you are walking among are perhaps 12 to 15 feet high, in the wild such trees can reach more than 100 feet. To your left is one of the Garden's mahogany trees, *Swietenia mahagoni*. This is a small specimen of one of the larger trees of the Latin American rain forest, and it is the tree Europeans in the eighteenth and nineteenth centuries pushed to the brink of extinction as they sought out its wood for beautiful furniture.

Because there is no winter in the tropics, the total growth of forests there each year may be three times that of the northeastern United States oak and maple forests. A single large rain forest tree can produce three pounds of pure sugar per day as well as growth so incredible it seems to occur almost before your very eyes.

If a Tree Falls in the Rain Forest...

As you approach a hanging curtain of vines, you sense that something very dramatic has happened. And indeed, such a scramble of climbing plants often is a sign that one of the great old trees such as the kapok has fallen in the forest. The force of storms, the saws of lumbermen, or the sheer weight of epiphytes—plants living on the branches and collecting heavy pools of water—may have contributed to the death of such adult trees. If you were within earshot, you would have heard a prolonged crashing noise—some

visitors to the forest characterized it as the sound of a locomotive tearing through a row of wood-frame houses.

The large branches arching over the walkway have fallen from a kapok tree, *Ceiba pentandra*. Before their fall, the branches were high in the canopy of the forest, and so they are encrusted with epiphytes such as bromeliads.

The death of a tree more than 100 feet tall

The kapok has fallen, and sunlight fills the gap.

and the domino effect of smaller ones falling suddenly create large openings in the canopy. Sunlight pours through the "gap" to the floor, encouraging germination of dormant seeds. New growth occurs rapidly there, often in the form of pioneer trees and plants designed to grow fast and fill the gap, while it lasts. The seeds of species in genera such as *Cecropia* and *Ochroma* wait buried in the soil until trees fall and clearings occur.

Later, as the *Cecropia* die out, making room for the seeds of longer-lived species to sprout, the gap is well on the way to regeneration. However, before the *Cecropia* die out, the seeds they produced in their short lives have dropped into the soil where they will await the next canopy gap. Scientists point out that such clearings, large or small, are critical for maintaining species richness in the rain forest. The gaps represent a natural disturbance that is often quite different from those occurring as a result of unwise, large-scale logging practices which, by increasing light, temperature, and resultant drying of the soil, destroy the fragile root mat, thus dealing a death blow to the forest.

New Therapies for Cancer and AIDS May Be Just Ahead...

Drugs derived from the rosy periwinkle provide effective treatments for childhood leukemia and Hodgkin's disease.

The gap in the rain forest is also the site of the Healer's House and Garden, and here you will see the rosy periwinkle, from which we derive a modern drug to fight certain cancers.

Whether for poisoning predators, attracting pollinators, or repelling competing species, many plants of the rain forest produce complex and powerful chemicals. Discovered, isolated, extracted, and studied by scientists such as those at the Garden—often with the help of traditional healers with whom the scientists work—these substances have historically formed the raw materials of humankind's pharmacy. Aspirin and codeine are among the best known plant-based drugs in the world today, and they were discovered by studying traditional folk medicine. Although scientists, in collaboration with local people, so far have been able to study only one percent of tropical rain forest

The pilocarpus tree is the source of pilocarpine, used in western medicine to treat glaucoma and recently approved to treat dry mouth syndrome. Depicted here is the species Pilocarpus pennatifolius.

MEDICINAL PLANTS

by Michael J. Balick, Ph.D.
Director and Philecology Curator, Institute of Economic Botany

For thousands of years, up to the present, plants have played an important role in human health. Ancient societies such as the Egyptians, Indians, and Chinese all have a rich history of using medicinal plants, and these uses have been codified in the earliest writings of these great peoples. For example, the Egyptians recorded their medical systems on rolls of papyrus, beginning some 4,000 years ago. The current, widely-held belief that all medicines are the product of complex chemical synthesis is incorrect, as more than 120 prescription drugs are obtained from higher plants (plants bearing flowers or cones), resulting in annual sales of over $12 billion in the United States. Many pharmaceutical companies are exploring the potential of the plant kingdom as a source of new medicines. This search is truly a race against time, as deforestation and species extinction, coupled with the loss of knowledge about traditional medicinal uses of plants, reduce this vast phytochemical reservoir with increasing speed. The maintenance of natural areas where so many medicinal plants are found is an important objective of conservation activities today, as billions of people around the world depend on the availability of these species for their health-care needs.

plants for their medicinal potential, 25 percent of all prescription drugs dispensed from pharmacies in the United States over the past several decades have contained active ingredients that have been extracted from plants.

Garden scientists are now working with pharmaceutical companies to discover new medicines, because the future of new pharmacological research remains in the rain forests, where hundreds of plant species—some not even named yet—possess anti-cancer and other beneficial properties.

The area of science known as ethnobotany concerns the whole range of relationships between plants and people, including how we use plants as a source of medicines, food, fuel, fiber, and fragrance, as well as for spiritual and religious pursuits. Ethnobotany relies on the knowledge of indigenous peoples for its insights. However, these people are disappearing, along with the plants. Tapping local botanical knowledge to address global health, food, and resource challenges is like running two races simultaneously against a relentless clock.

Don Eligio Pante taught Garden scientists about the medicinal plants of the Maya in Belize.

On just one of the nearly 1,000 field research expeditions undertaken by Garden scientists over the past century, ethnobotanist Brian Boom, Ph.D., lived for five months in a village of Chácobo Indians in Bolivia. Dr. Boom determined that the local people used 305—85 percent—of the 360 plant species he surveyed in the area; of these, 174—more than half—were boiled, ground, or otherwise rendered into medicines for every ailment and need from headache to snakebite, from rheumatism to contraception.

Above: Bananas.
Right: Sugar cane.

It is for more than medicines alone that rain forest plants hold great potential. With most of our food today coming from only 13 crop species such as rice, wheat, and corn, the supply—especially in poorer parts of the world—is dangerously vulnerable to crop pests, diseases, and the vagaries of climate. It is often in the species-rich tropics that new food resources are found. Garden scientists bring plant samples home from the field and then identify them through comparison with existing, named specimens in the Garden's herbarium, a collection of 6 million preserved plants, one of the greatest botanical databases in the world. Although the rain forest is the world's greatest natural pharmacy, its beneficial secrets for human beings have only begun to be unlocked.

The Second Biodiversity Crisis: A Shortage of Scientists to Do the Job?

In addition to conducting basic and applied research, the Garden also functions as a teacher. The more public crisis—the struggle to preserve the great species-rich areas of the world such as rain forests—masks a second or silent biodiversity crisis: namely, the severe shortage of trained botanists, ecologists, and land-use managers. For more than a century, the Garden and its sister institutions around the world have been a resource for training botanists, conservation officials, and land-use managers in

Michael Balick, Ph.D., with local healers in Belize.

the United States and elsewhere. By also helping to establish pioneering graduate research and training programs in Brazil and other countries, producing extensive publications, and providing additional research assistance, Garden scientists work with their colleagues to develop the scientific tools and knowledge that will ultimately help policymakers make decisions about the future of these spectacular ecosystems.

The Great Chain Reaction

As you continue on your journey through the Tropical Lowland Rain Forest, where you will see the great buttressed trunk of a giant kapok—one of the largest trees of the rain forest—remind yourself that the true marvel of this biome is not in any individual species or feature, no matter how spectacular or unusual. Rather, the grandeur is in the dynamic, often delicate balance among the myriad varieties of life that the rain forest contains. The rain forest is a plant-animal-soil-climate world unto itself, an ecosystem whose viability depends on our understanding, and then protecting, all its components.

As you walk up the Skywalk, imagine yourself moving through the layers of forest up from the understory, to the

canopy, and finally to the emergent trees that stick out like great green umbrellas. In the rain forest, the trees, seen from on high, form a tight weave, a metaphor for how all features of life are bound in the dynamic weave of the forest, where hundreds of species of plants and animals share the same few acres. Scientists have come to believe that conservation of the whole ecosystem is ultimately the chief way—maybe the only way—to conserve an individual species.

A kapok tree growing in the Conservatory for almost a century.

The Decision: Conservation

There is a huge amount at stake, and yet humankind's lack of understanding of the complexities of the rain forest is still profound. Of the total estimated 10 to 100 million plant and animal species on earth, only about 1.4 million, which includes an estimated 300,000 plants, have been described by systematic botanists. Of the remaining millions, many live fragile, crowded, interdependent lives in the rain forests. Who can say what the potential benefit of these species is to humankind?

The lowland tropics is a lithograph in Karl Friedrich Philipp von Martius' Flora Brasiliensis, *1840 - 1906, Vol. 1, Pt. 1. The LuEsther T. Mertz Library at The New York Botanical Garden.*

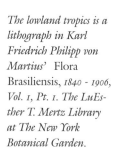

For example, in an area near Río Palenque, Ecuador, one of the most commonly used trees for house construction was only recently found to be a species new to science. Because rain forests are being destroyed at an alarming rate and many species cannot long survive on or near cleared crop land or plantations, extinction of the great heritage of our biodiversity is occurring even as you read these words.

Rain forest devastation in French Guiana.

When young Charles Darwin came to the Atlantic coastal forest of Brazil, the species he saw left him speechless. "Wonder, astonishment, and sublime devotion fill and elevate the mind," he wrote. Today, little more than a century and a half later, 95 to 99 percent of that particular tropical forest is gone. While extinction of species is a normal part of evolution, the extinction rate that occurs in the rain forest today is estimated to be 1,000 times greater than in pre-human times. Averting such unacceptable losses in the rain forests will be one of humankind's greatest challenges for the next century.

AQUATIC PLANTS

In the Aquatic Plants gallery, into which you've now stepped, pause for a moment and enjoy the fascinating plants in this tranquil and transitional space between the Conservatory's Tropical Lowland Rain Forest and the Tropical Upland Rain Forest biome where you will soon be traveling.

Papyrus

The Garden's collections include many specimens of papyrus, *Cyperus papyrus*, the plant made famous in history because, in the ancient world, its stem was "processed" to make paper. Papyrus specimens flank the entrance to this gallery.

Victoria amazonica

Many other plants, of course, live in even closer association with the water than does papyrus. They can be immersed in it, living their whole lives below the surface, or above but attached to the bottom of a pond such as the wonderful *Victoria amazonica* in the tropical pool in the courtyard of the Conservatory, which you can visit in warm weather. This giant water lily was a kind of poster

Facing: Cyperus papyrus *in the Aquatic Plants gallery.*

Victoria amazonica.

plant of the Victorian era. Named for the English queen after its discovery in the Amazon, *Victoria amazonica* was famously photographed. The photographs often showed a small child on the plant's outstretched pads, which can be more than three feet in diameter. Garden scientists discovered how the plant traps beetles by night in its flower, and as the beetles feed, they are covered in pollen grains that they carry to another flower when the *Victoria amazonica* releases them in the late afternoon.

Such aquatic plants have characteristics that keep them from getting waterlogged and expedite respiration. They also take advantage of each feature of their environment— for example, water lilies use the tensile strength of the water surface to bear the weight of their pads.

The passion flower is a vigorous climber.

Climbers—Twiners, Scramblers, Creepers

The vines you see growing on both sides of the gallery are very "smart" plants. Competing with trees in the crowded rain forests, vines don't waste time. Without quick growth, they die. Instead of expending their valuable sugar energy to manufacture trunks for support, vines—a type of climbing plant—just lean or creep or twine themselves or scramble over and around trees and shrubs, borrowing support from other plants' stems and trunks along the way.

The climber's greatest priority is to ascend toward the light and not to slip back once progress is made. To do this vines and lianas—lianas are woody vines usually found in rain forests—twine or climb by arm-like tendrils, hooks, adhesive pads, and even new roots, called adventitious roots, that grasp or adhere to the host plant as they ascend.

Charles Darwin, who in 1875 wrote the first book on the behavior of climbing plants, was fascinated with why the twining almost always takes place clockwise and rarely counterclockwise. What is your guess?

Many climbing plants are highly bioactive, which means they produce many chemical compounds, often full of resins or toxins, that discourage animals from eating them as they grow. These substances such as curare are of great interest to economic botanists and to healers both in the forest and at the modern research laboratory.

The fountain is the focal point of this gallery.

The fountain at the center of the gallery is a nineteenth-century French cast-iron sculpture, and the bluestone floor on which you are walking is very similar to what was in the Conservatory when it opened a century ago. That nineteenth-century gentleman, Charles Darwin, would have been very comfortable in this gallery, as we hope you are.

Before you enter the cool mistiness of the Tropical Upland Rain Forest biome, your next destination, remember that plants were initially aquatic—the algae—and that terrestrial plant life on which all animal and human life depends, emerged from the waters.

TROPICAL UPLAND RAIN FOREST

The heat of the lowland rain forest ebbs away as you enter the next biome, the Tropical Upland Rain Forest. The plant life in this biome represents that which occurs in rain forests at altitudes between 3,000 and 10,000 feet or more. Although they receive as much water as the lowland rain forests (which were below 3,000 feet), upland rain forests are engulfed in swirling mists of clouds formed by the cooler temperatures and the wind present at the higher elevations. Here, too, there is so little light that dense undergrowth on the floor of the forest often exists only in clearings and by stream beds.

As you walk, imagine yourself climbing down a narrow footpath that, in the actual forest, would be strewn with moist dead leaves. Mosses, ferns, and lichens abound and only begin to decline at the highest altitudes. The top of the forest canopy, which has also been dropping in height with elevation, is wreathed in clouds. You have now arrived in the cloud forest.

Among the most extraordinary upland rain forests are those in Mexico, Costa Rica, the Andes, and in the Guayana highland of northern South America. These are areas very well known to the scientists of The New York Botanical Garden.

Walking the steep mountain paths of the actual upland rain forest, you would notice, in particular, certain kinds of plants—orchids, mosses, and ferns—all growing in profusion and often clustering wherever a gap has opened. These are all adapted as intricately and interdependently as in the lowlands, but to this cooler, changing, and profoundly wet environment.

The air is misty and humid in the high mountain forests of Latin America.

A lithograph of an upland forest with bromeliads in Karl Friedrich Philipp von Martius' Flora Brasiliensis, 1840– 1906, Vol. 1, Pt. 1. The LuEsther T. Mertz Library at The New York Botanical Garden.

Aerial Gardens—The Epiphyte Way of Life

Because water is omnipresent in the air of the upland rain forests, not just in the ground, rooting on the forest floor is not the only way for a plant to live there. An extraordinary number of plants in these forests are epiphytic. The word derives from the Greek words, "epi" and "phyte," meaning, literally, "on plant." Epiphytes are not a species but a plant life-style adapted by many species.

Epiphytes live at least some part of their life cycle, and sometimes all of it, on the branches, trunks, roots, or leaves of other plants, and even on each other. Epiphytes are not parasites, which invade the body of the host plant to raid its food supply; rather, they strive for light, water, and nutrients while in residence—often on the upland rain forest's large trees where they are permanent botanical squatters.

While water is relatively easily absorbed from the air here, the sunlight essential to photosynthesis still is at a great premium at low and middle levels in the forest. Therefore, some epiphytic plants often live high in the branches of trees to gain access to the light without investing in lengthy stems. Scientists don't know how many plants live epiphytically, but the estimated number, particularly in the rain forests, is very large, including an extraordinary 20,000 epiphytic species of orchid.

Tillandsia lindenii *is a common epiphyte.*

Some epiphytes dangle down hairy roots to catch moisture from the air. Leaves of other epiphytes are sometimes arranged to funnel water to "storage tanks" at the center of the plant. In some larger epiphyte species of bromeliads such as *Vriesea imperialis* (one of the largest of the world's bromeliads), the leaves form great cones so that an entire

Vriesea imperialis.

gallon of water is often held in the tank at the center of the leaf spiral. This epiphytic burden puts stress on tree branches, and it is no wonder that the epiphytes at times wind up killing their hosts. Along with this water, minerals and any available organic matter, including dead insects and debris, accumulate in the plants, serving as reservoirs during dry periods. Insects, frogs, and other small creatures also reside or hide in the thick accumulated epiphytic tangle.

Curator of Orchids Keith Lloyd works with the Garden's major collection.

Make Room for the Orchids

These cool wet upland rain forests are home to one of the largest of all the families of flowering plants, the orchids, some of which you can see up close in the glass case at the base of the forest path. It is estimated that up to 30,000 orchid species grow worldwide, and they constitute nearly 10 percent of all species of flowering plants on the planet. Growing throughout the world, except in very cold and extremely dry climates, the majority are tropical epiphytes inhabiting the rain forests. Up to 50 different orchid species have been found growing on a single rain forest tree.

The profusion of orchid species and the evolution of orchids' elaborate structures are the result of adaptations made to gather water and nutrients and, in particular, to reproduce. For example, swollen stems, called pseudo-bulbs, store water, while the aerial roots absorb moisture from the air. Orchid flowers, in all their extraordinary variety of colors, aromas, and long languorous shapes—the inflorescences (groupings of flowers) of the dancing lady orchid, *Ocidum sphacelatum,* sometimes extend up to five feet—attract pollinators. A specific pollinator for each species, usually an insect and often a bee, likely accounts for the huge variation in the features and qualities of orchids. This close and parallel mutual adaptation, for example, between orchids and bees and other pollinators is also called co-evolution.

The orchid house near the path you have climbed contains some of the 5,000 orchids in the Garden's green-houses, one of the leading collections in the world. Cared for by Curator Keith Lloyd, the Garden's orchid collection includes more than 740 species. Ever since the first orchid nurseries were established in the mid-nineteenth century, collecting these exotically beautiful and sometimes bizarrely constructed flowers has been a craze and cultivating species to make hybrids has been an absorbing horticultural hobby for millions of people. There are now more than 60,000 orchid hybrids, with strict rules for registering and naming new variants. The Garden's

Due to its vibrant orange color, this orchid from Central America, Cattleya aurantiacea, *has been widely used in hybridizing.*

collection serves both as a display of beauty and a source of instruction. Just as importantly, the collection also preserves tropical species in genera, whose numbers are declining— such as *Cattleya* (the classic prom corsage), *Laelia,* and *Epidendrum*—due to habitat destruction and overzealous collecting.

Terrace Garden.

Fighting Deforestation of the Rain Forest

The Terrace Garden that rises into view to your right as you pass the orchids is symbolic of how human beings, resident in the rain forests for millennia, have developed profound knowledge of how to maximize use of available land. This ancient technique of farming on steep slopes retards erosion of the hillside.

An important area of research for Garden scientists like Christine Padoch, Ph.D., is the study of how people manage and use plant resources in the different climates and terrain of the tropics. She has

Christine Padoch, Ph.D., conducting research in the field, consults with Señor Vela, a local farmer, in Peruvian Amazonia.

View of a three-year-old cultivated field containing manioc (Manihot esculenta) *in the foreground, cashew* (Anacardium occidentale), *macambo* (Lonchocarpus nicou), *and peach palm* (Bactris gasipaes). *Agroforestry systems, because of their complexity, often are difficult to see as such, let alone understand.*

AGROFORESTRY

by Christine Padoch, Ph.D.
Curator, Institute of Economic Botany

Some Amazonian agroforestry systems are cyclical and are based on an alternation of intensively cultivated plots and subtly managed fallows. One such example was studied by a team of United States and Peruvian scientists on the Yaguasyacu River in the Amazon rain forest of South America. They found that the Bora Indians manage fields long after agricultural crops have given way to diverse forest gardens of planted and wild species.

been studying the river people in the Amazon rain forest of South America and learning about their extraordinary diversity of farming techniques in floodplains, which usually occur in lowlands, but are not restricted to the rain forests. In the village of Santa Rosa in Peru, for example, she documented 14 different types of agriculture that take advantage of the smallest environmental changes to grow crops such as rice on the mudflats for a brief two-month period each year. Growing crops in areas where even the government has said it was impossible, local people have productively and sustainably used the floodplain where the waters at flood stage regularly reach 30 feet above their low-water level.

Sometimes, ignorance of such suitable farming techniques on the part of some displaced and landless populations, who have been forced into the rain forest because of lack of land elsewhere, has resulted in deforestation. Likewise, large-scale logging, cattle ranching, and an excessive reliance on growing of single crops, which is called monoculture, often put rain forest ecosystems at risk. fighting deforestation of the rain forest is as complex a proposition as the biome's species are diverse. What works in one situation will fail elsewhere. Strategies for conservation are very site-specific.

Trunk of Cyathea princeps *collected in the wild by John Mickel, Ph.D.*

Ferns—Back to the Future

As you entered the Tropical Upland Rain Forest biome, in front of you was a handsome tree fern, *Cyathea princeps*, one of many such large ferns in the Garden's collections. John Mickel, Ph.D., the Garden's world-famous fern specialist, collected this particular specimen from the wild, in Mexico. Dr. Mickel brought many of the ferns in this exhibition to New York.

It is appropriate that you end your tour of the Conservatory's rain forest biome with a visit to the ferns near the waterfall at the far end of the gallery, because, with the exception of bacteria, algae, and possibly mosses, ferns and fern allies are the oldest living plants on earth. They date back at least 400 million years.

Unlike flowering plants and gymnosperms that reproduce by seeds, ferns have a special reproductive system that relies on spores. Spores are produced in tiny cases, called sporangia, on the back of each fern leaf. When released, they grow into a tiny heart-shaped plant that is capable of producing male and female sex cells which, when joined, create a new fern plant.

Most of the 12,000 known fern species live in the tropics, with large numbers in the upland rain forests. They range in size from the one-eighth-inch aquatic mosquito fern to huge tree ferns. The latter, in their growth to between 60 and 80 feet, evoke the prehistoric vegetation that was home to dinosaurs ages ago.

Some ferns, such as the bracken fern, *Pteridium aquilinum,* are pioneers in the retaking of landslides or man-made clearings. Other spore-bearing plants—mosses and liverworts—also grow in great profusion in the upland rain forest, literally holding together the often unstable soils of the mountain slopes, particularly in the ecosystems of the Andes.

Pteris altissima.

Over eons, the layers of now extinct fern allies have died, accumulated, compressed, and then hardened into coal. When we burn coal, we release the heat of these ancient sugars made through photosynthesis hundreds of millions of years ago.

As you leave the rain forests and enter a new biome, the deserts, the relationship of plants to people—the theme of your tour through the Conservatory—takes on another dimension. Not only are food, shelter, medicines, wealth, and the breath of life provided for us by the plants of today, but the plants of the deep past also sustain us.

DESERTS
OF THE WORLD

Most of the deserts of the world lie north or south of the equatorial rain forests you have just left, with shrublands and grasslands often intervening between forests and deserts. However, as you make the transition to the Deserts of the Americas and then to the Deserts of Africa, it is important to remember that the natural world is not a series of neat zones with precise borders.

Indeed, it is in the nature of plants to move, adapt, colonize, and invade every possible niche. For example, the desert cacti, those wonderful plants you are about to meet, have a few "cousin" genera—*Rhipsalis* and *Schlumbergera* cacti—that live as epiphytes in the rain forests. Although most of the 2,000 species of cactus live in the deserts of the Americas, a species of prickly pear, *Opuntia humifusa,* also grows wild over much of the eastern United States.

Documenting plant populations and deciphering the evolutionary links among them is the job of scientists such as the Garden's Noel Holmgren, Ph.D., and Pat Holmgren, Ph.D., who conduct research in the deserts and inter-mountain regions of the western United States.

Water, Water, Water...

Deserts often occur between mountain ranges where nature has created two high barriers to rain and air-borne moisture; an example is the Great Basin Desert of North America between the Rocky Mountains and the Sierra Nevada. The result is the defining condition of all deserts: minimal amounts of water, usually less than 10 inches per year in the dry deserts of the world and between 10 and 24 inches in the semi-arid deserts of the Americas and Africa. If the great botanical challenge in the rain forest is the fight

The Garden's double-trunked Boojum tree (Fouquieria columnaris) stands guard at the entrance to the Deserts of the Americas.

Noel H. Holmgren, Ph.D., is currently studying and classifying species of Castilleja, *or paintbrushes, for* Intermountain flora. *Being published by The New York Botanical Garden,* Intermountain flora *is a unique and comprehensive series that provides keys, descriptions, and illustrations of the vascular plants of the Intermountain region of the western United States. This region, delimited on scientific grounds, had been one of the largest gaps in the floristic coverage of the United States.*

FLORAS

by Noel H. Holmgren, Ph.D.
Mary Flagler Cary Curator of Botany, Institute of Systematic Botany

Scientists at The New York Botanical Garden have been among the most active in the world at producing "floras" (identification guides to plants of a specific geographical region) of the Americas. For example, current floras (13 volumes published; two forthcoming) describing the plants of the Northeastern, Northwestern, and Intermountain West regions of the United States were written by Garden scientists.

for light at the top of the canopy, in the desert the supreme struggle is for access to sufficient water for survival.

Mojave, Sahara, Gobi, and...Antarctica?

Deserts cover one-third of the earth's surface and yet are so harsh that barely five percent of the world's people live on them. Only some types of desert are sandy such as parts of the Sahara, and many have mountain ranges, basins, and huge canyons like the Grand Canyon. Deserts such as the Gobi Desert in Mongolia can also be extremely cold at night. They can even be in Antarctica, where frozen water is poorly available to plants and animals. From the point of view of the lichens and mosses growing there, Antarctica is, in effect, nearly waterless—a frozen desert.

Nevertheless, deserts are, despite myths to the contrary, profoundly alive with thousands of plant and animal species that have developed an extraordinary variety of adaptations to meet the dual challenges of survival in a desert ecosystem—protection from sunlight and acquisition and storage of water.

The Grand Canyon seen from Desert View on the south rim.

The desert fan palm, *Washingtonia filifera,* native to California, and a related species, the Mexican fan palm, *Washingtonia robusta,* which you can also visit in the Deserts of the Americas, have root systems enabling them to reach available underground water. The success of the well-known Joshua tree, *Yucca brevifolia,* which grows in elevated desert regions, is due to its tolerance for both drought and heat. And many plants of the desert are superbly adapted to take advantage of sudden rainfalls by germinating from seed, growing, flowering, and producing new seed very quickly.

Many of the desert plants you pass in raised beds as you move through this biome are succulents, the collective name for more than 10,000 thick-fleshed species designed for water storage. Succulents survive and even flourish in the desert and other environments through slow growth and a remarkable array of adaptations centered primarily on retaining astonishing amounts of moisture in roots, stems, or leaves. Among these succulents are many species of euphorbias, agaves, aloes, and sedums. Their shapes and sizes vary from clumps to rosettes of overlapping leaves

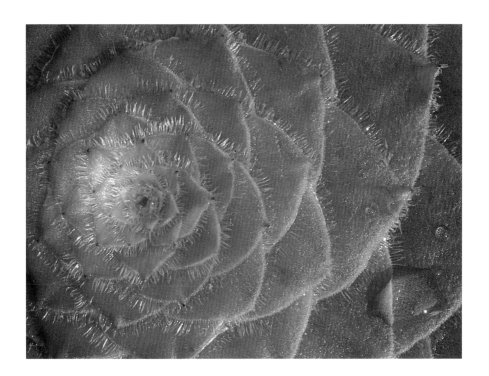

worthy of a jeweler's design. The most famous of the succulents are the cacti.

The overlapping leaves of the succulent Aeonium tabuliforma.

The Remarkably Engineered Cacti

Cacti can be startling in their often bizarre geometric shapes. But the appearance of cacti results from a unique evolution in which plant form, function, and survival are seamlessly interwoven. They adapted to minimize water loss. Cacti range in size and shape from the huge columns of *Pachycereus pringlei,* which can grow to 60 feet high and has branches shaped like candelabra, and *Dendrocereus nidiflorus,* with its massive trunk diameter, to tiny globular shapes surviving in clusters on the desert floor. The vertical "ribbing" on many cacti enables the stem to expand like an accordion when rainfall is plentiful; stored water may form 90 percent—or even more—of the weight of a plant.

While other desert plants have normal leaves, the modified leaves of cacti—the bristling spines—discourage predatory animals, radiate heat, and also efficiently capture dew

Lemairecereus stellatus.

and moisture in the air, which then drop to the ground to be absorbed by the cactus roots. If cactus leaves had not evolved over time into spines, expansive leaf surfaces would have lost moisture and eventually been burned by the killing sun. Therefore, cacti, in one of evolution's most remarkable engineering stories, also switched the site of life-sustaining photosynthesis from exposed leaves to fleshy stems. And that's not all. The pores, or stomata, of many species of cacti and other succulents, which in other plants open by day to take in carbon dioxide for photosynthesis, do so at night to minimize water loss.

Saguaro cacti,
Carnegiea gigantea,
in Arizona.

Saguaro Cactus, Keystone Plant of the Sonoran Desert

Among the grandest plants of the Americas, the saguaro, *Carnegiea gigantea,* dominates the landscape of the Sonoran Desert, which encompasses parts of northern Mexico and the southwestern United States. The saguaro is a giant that first grows branches, or "arms," when it is between 50 and 70 years old, and it can live to be 200. It may be 50 feet tall, extend its roots laterally 100 feet, and weigh as much as 14,000 pounds, most of the weight being precious water.

The saguaro's importance is not in statistics. Rather, like the keystone that keeps all the other stones in an archway from falling, the saguaro is the hub of the wheel in the desert ecosystem in which it lives. Woodpeckers make the many holes you might notice in the stem—for nests high out of the reach of predators. When the young woodpeckers leave, other species of bird such as elf owls and sparrow hawks enter, followed by spiders, insects, and bats. Below, near the base, mice and lizards burrow for moisture.

Like nature's cooperative apartment building in the desert, this micro-habitat around the saguaro is unique. In an environment with scant resources, the impact of this remarkable plant, which flowers only once every five years, is enormous; indigenous people eat the saguaro's fruit and, after it dies, use the skeleton for furniture or shelter.

136 Degrees Fahrenheit—in the Shade!

As you step into the Deserts of Africa portion of the biome, contrast the older deserts of Africa, Arabia, and Australia with the geologically younger deserts of the Americas, which contain mountains, basins, and flats. The former are full of undulating plains and vast expanses of soil that has been ground down over time from rocks into sand, reflecting their more ancient origins 100 to 200 million years ago. And they are hot! Azizia, Libya, recorded the highest official temperature in the shade—136 degrees Fahrenheit in September 1922.

While harsher areas have little botanical life, deserts such as the Namib, Karoo, and the Kalahari in southern Africa are far more lush and support many more plant species. Southern African deserts are also home to many aloes, euphorbias, and other succulents, which you may visit in the Conservatory's biome. Distinctive species of plants known as lithops, or "living stones," often grow in open

Lithops propagated from seeds collected in South Africa by the former Curator of Desert Plants.

sand-free areas, for example, among the pebbles in the Namib Desert of southwestern Africa. They store water in their two thick moisture-laden leaves. In severe drought, they are protected not only from extreme water loss, but their coloration and pebble-like shapes also provide excellent camouflage against thirsty herbivores.

Many species of African euphorbias that you can examine in this biome resemble cacti from the Americas, but they are not cacti. Like cacti, euphorbias have developed a tough juicy stem often ribbed so that it can expand, accordion-like, to hold water, and both often have modified their leaves into formidable spines. However, from an evolutionary point of view, euphorbias are unrelated to cacti, as seen most clearly when they flower. This process by which unrelated plants develop similar appearances under similar environmental conditions is called convergent evolution.

The Fragile Deserts

Deserts are expanding worldwide, particularly in places like the Sahel, south of the African Sahara. Unlike in rain forests, where biodiversity is richest, fewer species have evolved in deserts. Should climate change or exploitative human intervention disrupt the natural system, the limited amounts of water and other resources in deserts make quick regrowth and recovery less likely.

When deserts are irrigated to grow crops, the water brought in can often destroy more than it revitalizes, because water evaporating from soil pulls salts from the soil and deposits them on the surface. Without trees or other shade to protect the soil, the water evaporates rapidly, leaving the salt behind. Low rainfall prevents the salt from being washed out of the soil. Over time, the salt accumulates in the soil, rendering it unsuitable for agriculture.

When trees in the African savanna or grassland at the margins of deserts are cut for timber, or vegetation is overgrazed, further desertification follows. The kind of desert created in this manner is not nearly as diverse in species as a "natural" desert, is more subject to erosion, and is far more barren. Vast areas of low rainfall can best be conserved by documenting and understanding the existing relationships among plants, animals, and people, and then by maintaining these interactions through appropriate land management.

Lithograph of a desert scene depicted by A.H. Campbell for U.S. Pacific Railroad Exp. and Survey Reports, 1855–1860. The LuEsther T. Mertz Library at The New York Botanical Garden.

Aloe picitifolia in the Deserts of Africa gallery.

SPECIAL
COLLECTIONS

Horticulture at the Garden

Before you explore the world of subtropical plants and the Conservatory's dazzling seasonal exhibitions in the final leg of your journey, pause here a moment.

The art and science of cultivating and maintaining the plants you have seen in the biomes—and the flowers you now see all about you—is called horticulture. The New York Botanical Garden is a grand collection of natural landscapes, historic and contemporary gardens, and living plant collections from both familiar and exotic places. The Garden is dedicated to your enjoying the natural world as well as learning about it. In the relationship between plants and people, knowledge and pleasure should go hand in hand.

Therefore, if your visit is a bit of a renewal, a reconnection with the spirit from which all life derives, then you can thank the Garden's staff of horticulturists, who grow and develop different varieties—called cultivars—and care for the collections. More than 40,000 living plant species have been gathered and curated at the Garden throughout its 100-year history. Every day the horticulture staff cares for the Garden's 250-acre site, including a 40-acre tract of the last untouched forest in New York City and many specialty gardens such as those for roses and native plants.

Horticulture as a profession calls for a breadth of practice and expertise. It demands knowledge of subjects ranging from fertilizers to genetics. It also requires extraordinary patience to undertake work such as making 700 attempts to cross two varieties of daylilies. But it also involves the loving touch of a nurse, which, by the way,

Curator of Tropical Plants Francisca Coelho tends bromeliads in branches of the fallen kapok.

is also the profession of one of the Garden's current horticulturists.

Hanging Gardens and Carnivorous Plants

Above you are some of the results of the horticulturists' efforts. Often there are hanging baskets overflowing with the red, yellow, and orange flowers from the Garden's collection of begonias, a very large and varied genus. There are frequently many beautiful hanging baskets of ferns.

The carnivorous plants on display in the glass vitrine in the gallery include *Nepenthes,* popularly known as pitcher plants and trappers, whose

Davallia ferns in hang-ing baskets.

most famous species is the Venus flytrap. There are also flypaper plants and the aquatic insect-hunters, the bladder-worts. Most live in wet, swampy areas such as the Gulf Coast of the United States and, scientists think, evolved to be able to absorb insects to extract nitrogen and other nutrients missing from the plants' native boggy soils.

Insects slide down the slippery sides of the pitcher-like leaves of the *Nepenthes,* with their downward-pointing hairs, becoming mired in the leaves' sticky liquid. There they drown and are digested. Active insect trappers, such as the Venus flytrap in the genus *Dionaea,* snap their leaf halves shut when certain trigger hairs are brushed, trap-ping the prey. Sundews are passive catchers, covered with sticky secretions that only get more so as the prey struggles, causing the sundew's "tentacles" to curl over it for the final act of the digestive drama.

By isolating the enzymes that the plant carnivores produced and by demonstrating their likeness to those of animals, Charles Darwin proved that plants digest animal tissue. Subsequently, hundreds of species of carnivorous plants have been discovered by scientists and collectors fascinated with the seeming reversal of the pattern we expect in nature.

All of the plants exhibited in this gallery, many often extraordinarily challenging to grow such as the bonsai varieties, are brought in when ready from the Garden's Propagation Range, where they are grown. The Propagation Range is a two-acre network of working, behind-the-scenes greenhouses where Garden horticulturists care for, hybridize (that is, cross with other varieties), and prepare thousands of plants for display here and in the seasonal and subtropical exhibition galleries.

Here are two of specimens of carnivorous plants from the Garden's extensive collection, Sarracenia venosa, *above, and* Cephalotus follicularis, *below.*

SEASONAL EXHIBITIONS AND SUBTROPICAL PLANTS

Now step into the last of the Conservatory's galleries, where the seasonal plant displays are the centerpiece. These exhibitions, whose themes vary throughout the year, are among the most spectacular and well-attended horticultural events in the Northeast. They are a source of beauty and rejuvenation for city-dwellers and an inspiration for amateur horticulturists and gardeners, whose hobby, year in and year out, remains the most popular leisure-time activity in the United States.

In the winter months, particularly January and February, you will see daffodils, tulips, hyacinths, and other bulbous plants that are a much anticipated foretaste of spring. These bulbs have been "forced," or brought into bloom at the Propagation Range well ahead of their natural flowering time. A bulb is a thickened underground structure, mostly made of swollen leaf bases, which usually rests during part of the year. In horticulture, "forcing" is the demanding art and practice of inducing plants to flower a season earlier than the natural one.

In late spring, the bulbs are replaced by colorful annual and perennial flowers, herbs, trees, and shrubs. A perennial plant is one that produces new growth from a persisting part and lives year after year; an annual goes through its life cycle—it grows from seed, blooms, fruits, and dies—in the course of one year or less.

Flowers in the Spring Flower Exhibition are forced by the Garden's horticulturalists.

Spring in the Conservatory.

A cruciform house prepared for the Holiday Show.

In summer and fall, the displays feature mainly tropical plants such as the anthuriums. Also very notable are the Garden's chrysanthemums, including the Japanese kiku types. (The chrysanthemum is Japan's national flower and the imperial emblem.) And during the holiday season, the Garden's popular exhibit of large model trains is set among a display of the well-known holiday plants such as the

The Holiday Garden Railways Exhibition.

poinsettia, *Euphorbia pulcherrima*, and classic evergreen shrubs and trees. These are so named, of course, because they retain their green growth for more than one season. They include not only firs and pine trees but also broad-leaved evergreen plants such as holly, ivy, and rhododendron.

Plants of the Subtropics

Flanking the seasonal displays in the corner gallery are permanent exhibits of plants from the moist subtropics, mostly from eastern Asia. In the longer gallery, leading back to the Palms of the Americas gallery, are plants from dry subtropical areas with "Mediterranean" climates, including plants from Australia, South Africa, and California.

The subtropics lie between the tropics and the cooler temperate zones (such as our own) and are characterized by long growing seasons, minimal freezing weather, and a cool rainy season, conducive to sudden blossoming and growth.

The Wet Subtropics

Plants of the moist or wet subtropics occur where there is a warm, moist summer and a drier (or less moist) winter. Among these plants are the camphor tree, *Cinnamomum camphora,* the litchi tree, *Litchi chinensis,* and, most famous of all, the camellias. A genus comprising 80 species, growing in shade from 8-foot shrubs to 50-foot trees, the camellia has blooms that are among the most spectacular that nature, assisted by human horticultural ingenuity, has thus far created. There are now more than 3,000 named camellia varieties, of which the most widely known is the *Camellia japonica,* with its showy red, white, and pink flowers.

A tea plant.

The Boston Camellia sinensis *Party*

Originating around Chinese temples as many as 2,000 years ago, the tea plant, *Camellia sinensis,* grows leaves from which tea is derived. Associated with the sacred rites of temples, tea, when imported to Europe in the seventeenth century, also took on ceremonial qualities. In England, for example, tea was so prized it was kept in locked silver boxes, and afficionados evolved their own indigenous rites, such as "tea time" and "high tea." By the time of the American Revolution, the cultural and economic importance of tea had become so significant that outraged colonists, throwing British tea into Boston Harbor, were contributing to more than a political revolution. A cultural and gastronomic shift was being proclaimed as well: from consumption of Asian tea (on which the British maintained a lucrative monopoly) to the drink derived from beans of the *Coffea* plant (which was more easily obtained from the American tropics).

Plants of the Dry Subtropics—Australia, South Africa, California

Although geographically isolated from each other, southwestern and southern Australia, the Cape region of South Africa, coastal California, and Mediterranean lands have amazing similarities in their subtropical climate: dry summer, a brief but intense rainy season in winter, and a natural fire season.

As a consequence, remarkably similar vegetation grows independently in these separate locales. Low, scruffy evergreen bushes with leathery drought-resistant leaves, many of which flower spectacularly after the rains, dominate the

Unripened coffee beans.

An engraving in Matthew Flinders' A Voyage to Terra Australis, *1814.The LuEsther T. Mertz Library at The New York Botanical Garden.*

This olive tree remained in place during the restoration of the Enid A. Haupt Conservatory between 1993 and 1997.

universal botanical landscape in these subtropical places.

In the longer gallery, *Eucalyptus globulus,* a member of one of the most widespread and important plant groups of Australia, with over 600 species, dominates. Some eucalypts, known as gum trees because of their characteristic sap, have leaves and bark long valued as medicines to cure colds and sore throats. *Eucalyptus globulus* can reach 400 feet high.

Also in this gallery is the Garden's olive tree, *Olea europaea.*

The Refining and Life-Giving Subtropical Fires

When the brush fires, a natural feature of long, dry summers, break out, resins and oils common in the wood and leaves of eucalypts and other subtropical species fuel the flames. The fires are often so widespread that they are visible miles away. Many eucalypts such as the smaller Malleé ones have new-growth zones in tubers attached to their roots. Heat stimulates the tubers to release starches and to grow. Such subtropical regions often bloom luxuriantly after the fires and could not do so without such stimulation.

In a small corner of South Africa is a unique subtropical area where the dominant vegetation, known as fynbos, is home to 7,000 species of plants. Here, too, periodic fires create and maintain rich botanical life because they prevent trees from growing too large and crowding out the smaller plants. Among the popular garden plants and flowers that originated among the fynbos are species of *Pelargonium,*

Protea.

including tender geraniums. The fynbos also brought forth calla lilies, gladioli, gazanias, and species of *Protea*, many of which will germinate only after the heat of fire has cracked the hard seed coat. Within the Ericaceae heath plants there has been an explosion of species in the fynbos, with some 650 unique to the area, making the biodiversity of this vegetation type among the richest in the world. You can see some of these plants in the beds to your left as you proceed through the gallery.

The climate and soil conditions in California, particularly around Los Angeles, San Diego, and San Francisco, are also similar to those of the Mediterranean, with perhaps only 1 or 2 inches of rain of the total 10 to 20 falling during the arid summer months. Here, where water rationing even for household needs is more and more common, an increasingly popular approach to domestic landscaping in dry regions, called xeriscaping, has developed. Xeriscaping not only relies on traditional plants adapted to growing with limited water such as the cacti and other succulents you met in the desert biomes, but also embraces the full range of indigenous and drought-resistant species from the arid and semi-arid regions of the world. These often require no more water than that which naturally falls as rain and dew. Among these drought-tolerant species you may visit in this last gallery of the Conservatory is the California laurel, *Umbellularia californica,* whose leaves are used as a substitute for bay leaves—a popular herb.

EVOLUTION OF A GREAT BUILDING

The gardeners opened the doors and presto! The inside is
a thing of beauty and a joy forever filled with beautiful
and rare plants...

> —*from* The New York Times, *August 19, 1900,*
> *four weeks after the opening of the Conservatory*

The Enid A. Haupt Conservatory under construction in 1899. Historic photograph from the LuEsther T. Mertz Library at The New York Botanical Garden.

Just as the world of plants has changed over time, so too has the Conservatory building through which you have been traveling. Patterned after Victorian glasshouses such as those in England's Royal Botanic Gardens at Kew, the Conservatory was designed in a modernized Italian Renaissance style and built in 1899 by Lord and Burnham, the premier greenhouse designers of the era. Perhaps the most historic and beautiful glasshouse left in America, the Conservatory is an officially designated New York City Landmark and, along with the entire Garden—of which it is the internationally recognized symbol—it has also received the highest federal designation as a National Historic Landmark.

All the glass in the Conservatory was replaced in the restoration.

With more than 17,000 panes of glass covering nearly an acre of plants, the Conservatory is a unique and fragile building that, from the day it opened to the public in 1900, has required continual maintenance, repair, and renovation. Over the years, leaks developed, portions of the iron framing corroded, lightning struck, great snows weighed down and shattered glass panes, and

even once, in 1916, a large ring-tailed cat escaped from the Bronx Zoo, successfully invaded, and found refuge atop a *Euterpe* palm.

Beginning with the renovations of 1938 and 1956, the well-meaning human hand has also removed original features of the building, including the great ornamental entranceways. By the 1970s, the effects of nature, time, and a number of municipal fiscal crises had left the Conservatory in need of such fundamental repair that a choice had to be made between demolition and a full-blown restoration of a magnitude that had never before been attempted.

Enter Enid A. Haupt, a philanthropist and avid horticulturist whose gifts in 1978 rescued the building from certain collapse. Her love of the Conservatory and what it stands for in many ways has mirrored the botanical ardor, civic pride, and vision of the first director, Dr. Nathaniel Lord Britton, and the original group of philanthropists and municipal leaders who established the Garden a little more than a century ago.

With the New York City Department of Cultural Affairs, the Department of General Services (now the

Tarpaulins enshrouded the Palm House in 1994 to protect visitors as layers of lead paint were removed from its superstructure.

Department of Design and Construction), the Bronx Borough President's Office, and Mrs. Haupt pointing the way, the Garden began planning in earnest in 1989 for the most comprehensive restoration ever of the Conservatory. Under the leadership of Garden staff and the world-renowned restoration architect, John Belle, partner in the firm of Beyer Blinder Belle, the Garden gathered a unique and brilliant international restoration team—structural and preservation architects; exhibition designers; specialists in glazing and in the replacement of old metals with aluminum; experts in the computerization of building functions—to match the Conservatory's significance and stature.

As the restoration proceeded, in compliance with the New York City Landmarks Law and under the guidance of the New York City Landmarks Preservation Commission, the Conservatory received new drainage, electrical, mechanical, misting, and heating systems. Experts around the world in the replacement of historical materials assisted as the restoration team painstakingly replaced thousands of warped and aging glazing bars and window frames with new aluminum ones that replicate the style, color, and profile of the originals.

Between 1993 and 1997, through the generosity of Mrs. Haupt and The City of New York, the restoration was completed. It combines the best of the old with the best of the new, so that the Conservatory appears today much as it

did when the very first visitors lined up in their Sunday best to see what the scientists and horticulturists had wrought. Only now, vents and windows close and open at the touch of a finger on a keyboard, a fiber-optic cable runs underground beneath the rain forest and desert biomes the full length of the building, and the most sophisticated, state-of-the-art systems and equipment are contained within a dynamically restored historic structure.

As a consequence, the urgent modern scientific insights of conservation, ecology, and biodiversity—some of which you have learned about in your walk through the Conservatory—are now being communicated to a wide audience in a computer-controlled "crystal palace" that will remain both exquisite and structurally secure for generations to come.

The Enid A. Haupt Conservatory in 1997, after a comprehensive four-year restoration.

OUT INTO THE
WORLD OF PLANTS

The T. H. Everett Rock Garden on the grounds of The New York Botanical Garden in early spring.

You have now journeyed through rain forest and desert biomes, aquatic and subtropical zones. In all, you have visited approximately 3,000 plants that live in this great glasshouse, including both descendants of historical collections that have been here since 1900 and more than 1,000 new plants assembled for this new era at the Conservatory.

Outside are many more plants to visit, those that can survive outside a greenhouse, in the temperate zone that we share with them. Everyone at The New York Botanical Garden hopes that visiting here will open up a whole new

world for you, because beyond the Garden are hundreds of *The world outside!*
thousands of other species, many of which are waiting to
be discovered, described, understood, and perhaps used.
Maybe you will travel to these forests, savannas, deserts,
and mountainsides and even undertake the increasingly
vital work of understanding the relationship between
plants and people. In understanding and conserving the
plant world, we do no less for ourselves.

About This Book

This is the first volume of a two-part set of guidebooks about The New York Botanical Garden to be published in 1997 and 1998. The second volume will be a guide to the parts and plants of the Garden outside the Conservatory—the temperate collections.

This book has been created concomitantly with the completion of the Conservatory and the replanting of the collections in late 1996. The text is intended to interpret tropical, subtropical, and desert plant collections and exhibitions for the public from a general, humanistic point of view. We believe that all educated citizens, especially if they do not have a background in biology or the natural sciences, need and want to understand more about the plants. Allan Appel's writing was supported by the research of Paxton Barnes.

The photography is principally by Christine M. Douglas. The pictures were taken in late 1996 and early 1997, in some cases only a few days after the collections were replanted. As time passes, the plants grow, and the exhibitions mature, these photographs will document the early days.

The book was designed by Abraham Brewster and Ron Gordon of the Oliphant Press. The typefaces used are Galliard and Mantinia.

This publication was made possible, in part, by a grant to The New York Botanical Garden by Furthermore, the publication program of the J. M. Kaplan Fund.

Photographic Credits

Michael Balick, Ph.D. *pages 27, 43, 45*
Adam Bartos *page 87*
Tori Butt *page 72*
Carol Gracie *pages 17, 26, 38, 39, 47*
Mick Hales *page 10*
Andrew Henderson, Ph.D. *pages 20 (bottom), 23, 25, 29*
Robert Phillips *page 7*
Allen Rokach *pages 56, 58 (bottom), 78 (top)*
Marcia Stevens *page 76*

All photographs, except for those listed above,
were taken by Christine M. Douglas.